Showdown

Real or Pretend

Dona Herweck Rice

Publishing Credits

Rachelle Cracchiolo, M.S.Ed., *Publisher*
Conni Medina, M.A.Ed., *Managing Editor*
Nika Fabienke, Ed.D., *Series Developer*
June Kikuchi, *Content Director*
John Leach, *Assistant Editor*
Kevin Pham, *Graphic Designer*

TIME For Kids and the TIME For Kids logo are registered trademarks of TIME Inc.
Used under license.

Image Credits: All images from iStock and/or Shutterstock.

Library of Congress Cataloging-in-Publication Data

Names: Rice, Dona, author.
Title: Showdown : real or pretend / Dona Herweck Rice.
Description: Huntington Beach, CA : Teacher Created Materials, [2018] |
    Audience: K to Grade 3.
Identifiers: LCCN 2017029998 (print) | LCCN 2017031618 (ebook) | ISBN
    9781425853242 (eBook) | ISBN 9781425849504 (pbk.)
Subjects: LCSH: Animals--Juvenile literature. | Animals, Mythical--Juvenile
    literature.
Classification: LCC QL49 (ebook) | LCC QL49 .R483 2018 (print) | DDC 591--dc23
LC record available at https://lccn.loc.gov/2017029998

Teacher Created Materials
5301 Oceanus Drive
Huntington Beach, CA 92649-1030
http://www.tcmpub.com
ISBN 978-1-4258-4950-4

Some animals are
real.
Some are pretend.

A horse is real.

It runs.

Pegasus is
pretend.

It flies.

A crocodile is real.

It breathes air.

A dragon is
pretend.

It breathes fire.

Real or pretend?